ARMY TWITTER:

The good, the bad, and the ugly

Yamilet Rivera

Gautam Jain

Copyright © 2019 by Yamilet Rivera. All Rights Reserved. No part of this book may be reproduced, or transmitted in any form or by any means, including photocopying, recording, or other electronic or mechanical methods, without prior written permission of the author, except in the brief quotations embodied in critical reviews and certain other noncommercial uses permitted by copyright law.

Copyright © 2019 Yamilet Rivera

All Rights reserved.

ISBN: 9781710749571

Content

Acknowledgement 6
INTRODUCTION 7
THE GOOD 10
    Friendship 11
    Teamwork makes the dream work 14
    One in an Army 15
    News, News, News 16
    Studying? On twitter? 18
    Theories 20
    I see an intellectual 22
THE BAD 24
    Following the wrong account 25
    Account purposes 27
    Unpopular Opinion 34
THE UGLY 37
    Mental health 38
    Scammers 40
    Fan wars 43
    Misogyny, Xenophobia and Racism ... 45
CONCLUSION 52

Disclaimer	54
About Authors	55
References	57

Dedicated to the best fandom ever, my understanding husband and irreverent daughter- Yamilet Rivera

Acknowledgement

First of all, to BTS as without them, we wouldn't be experiencing this rollercoaster. Second of all, to all the ARMYs that makes this constellation brighter than a million stars.

INTRODUCTION

This guide's purpose is to help BTS fans, called ARMYs, to navigate Twitter universe. It is very common to link Twitter with the most toxic behaviour a human can portray online, because of the anonymity that is easy to obtain. In case a non-ARMY or an occasional listener is reading this book, BTS or Bangtan Sonyeondan (방탄소년단) is a Korean group that is the is the most successful, biggest boy band in the planet. According to Billboard, BTS has had three number one albums by 2019 in the Hot 200 in less than a year[1], a record that was set in 1996 by The Beatles. By November 2019, Map of the Soul: Persona is the best-selling album worldwide according to United World Chart[2]. These are two of the most impressive records BTS has had this year. Although they have 22.6 million followers, they are one of the accounts with most engagements (let's compare the 110.2 million followers Barack Obama has). Recently, they have been achieving a million likes less than 24 hours for every post, and they have won Billboard's Top Social Artist for three years in a row[3]. Currently, they are at #1 after 151 weeks. Also, they hold the record for the most viewed YouTube video in 24 hours with 74 million streams. Having said that, non-ARMY/occasional listener, you are welcome.

All these records could not be possible if it weren't for the Adorable

Representative MC of Youth (ARMY) that have been beside BTS since debut in 2013. There are other platforms where the group posts updates, videos, news, etc., but it is Twitter the source of it all. It is twitter the platform that can build or destroy. This guide will help you how to build a community, to be part of it, and to avoid unnecessary drama.

THE GOOD

I have been on Stan twitter for two years now, and I have met a lot of interesting people throughout the course of this period. I found that, though Twitter itself is a very scary place; there were some good things about it.

Friendship

Like I said, I am still there as a Stan or a fan Twitter Account. For the ones that do not know the concept, it is a profile solely to follow a certain famous person or group of people. I made the decision early on, because I know my in-real-life friends would not want to see a bunch of re-tweets, quotes, or even likes from BTS.

Through this tight-knit community that I am immersed in, I have found good friends online. They are almost my age, and, of course, we keep each other updated of the latest news, either good or bad. About the group, it is very easy to know firsthand tour dates, album or single releases, online promotions for their merchandise, which apps to use when a live broadcast is coming, how to vote for different awards, which awards are currently on and their due dates, lyric transcriptions and explanations which I found extremely useful since I do not know Korean.

In 2019 my daughter and I were blessed of being able to attend a BTS concert at MetLife Stadium, New Jersey on May 19th. There was an account I contacted some ARMYs that scheduled a meeting before the concert, so we could get in the mood. I wasn't able to attend at the moment, because of tight schedule between the bus arrival and the actual concert. Still, a lot of adults got together, and they found each other through this twitter account.

This was also a great way to continue with projects for the concerts as all dates had a different account for different projects for the boys and ARMYs.

I would also like to acknowledge that I have found not only online friends, but in real life, too in the most peculiar ways. One of them was in a competition, – I was the moderator of the participant group - and she opens her laptop. I could not believe the BTS wallpaper in all its glory. Another one shared an account of a k-drama application. Guess what her password was? *'A member's name'* It is the power of ARMY.

Teamwork makes the dream work

This catchy phrase was posted for the first time on March 19, 2013 even before there was an official debut, June 13th that same year. It not only refers to their work ethic, but also how ARMYs conduct either in their personal in-real-life, but also when it comes to work together as a fandom.

There are three important accounts, in my opinion, which are the backbone of what BTS philosophy is.

One in an Army

This account was created in 2018, five months after BTS launched LOVE MYSELF #ENDViolence campaign along UNICEF. According to their website, Ann (username @rwapmon) calls out ARMYs if they wanted to help for a Syrian relief project[4]. The account was born, and for one year and three months (June 2019), ARMYs had raised USD$46,000 for different projects around the world. Just to name a few: six semesters of college scholarships for orphans, 6,000 meals for at-risk-kids in Rwanda, Rebuilding

Gili Islands after an earthquake that happened on August, 2018, one year of basic foods for LGBTQ refugees, among others.

This is the perfect way to help other and get involve with the community *'Global and local'*.

News, News, News

Even though this account only has almost half a million followers, compared to the 22.5 millions (as of November, 2019) BTS has, it is impressive how one account dedicated solely to update BTS chart positions around the world @btschartdata was able to touch the fandom's heart on November 2018, after a Japanese scandal[5] and encourage us to buy 2!3! (The first song BTS dedicated to ARMY) and in matter of hours, it peaked at #4 on Worldwide iTunes[6], and had 35 #1s around the world[7]. This was the first time that we saw the impact of our emotional state into the music industry. This impact happened with a two year-old song.

The second important account with 1.8 million followers, @btsanalytics, focuses on news, voting and comeback guidance. Nowadays, it is hard to keep up with a lot of award polls, so this account generates images, gifs or short videos to explain how to vote and when to vote. Also, most importantly, it explains what the objectives of every comeback are. For example, the last one for 'Map of The Soul:

PERSONA', one objective[8] was to be part of the Top 10 HOT 100 on Billboard. It was achieved with the single "Boy with Luv ft. Halsey" on #8[9].

I have to say that these are not the only accounts devoted to charts. There are others specifically for each region or country. There are others that focus only on votes, or news, or comebacks. Now, the main point here is the way ARMYs are aligned and have a common ground and objectives. Also it is remarkable how they trust a lot of accounts that are determined to make BTS untouchable by the ones that don't want them to succeed.

Studying? On **twitter?**

Yes, believe it or not, there is an account, @armyacademics, committed to help ARMYs academically in six different areas: math, languages (Korean, English, Spanish, and French), science, social and computer science, art, and tips & methods to improve student life. Each area has its own twitter handle to organize tutors and students. I was a tutor for two months at the language academy teaching English. It was so much fun, because you can help others from across the world. Also, if you don't master a subject, there will be a fellow ARMY ready to help you.

Like I said before, it really is an awesome community.

Theories

Baby ARMY, this section is specially for you. Bangtan not only feeds us with albums, music videos, V-lives, but there is also another great pillar in BTS's success, and that is Bangtan Universe (BU). It is a parallel story that started with the album "The Most Beautiful Moment in Life" in 2016 up to now with Maps of the Soul: Persona in 2019. Big Hit created characters to represent the story of seven high school friends, one for each BTS member. The story is developed in a series of music videos, short films, notes that were included in albums, highlight reels, teasers, even a webtoon. Bang Shi Hyuk, Co-CEO of Big Hit and BTS's main producer, had the idea of creating a story that spoke to the target audience, but the response was so positive that he along with a team expanded the universe[10]. The fandom go nuts every time there is a new music or video release. Anything can be a clue for the next album, or for the next development of the story. There a lot of ARMYs are theorists at heart, but there is one that stands out of the bunch. Her alias is xCeleste on Twitter and YouTube. She created a half an hour video[11] to explain the BU's timeline and theories. Of course, everyone can participate, so put on a tinfoil hat and be ready to have a headache.

I see an intellectual

Ursula K. Le Guin, James R. Doty, Hermann Hesse, Alan Garner, Haruki Murakami, Antoine de Saint-Exupéry, Carl Jung. If you do not know who they are, you are in for a surprise. All these authors, ones more famous than others, have inspired BTS in one way or the other for their songs and albums.

The latest album Map of the Soul: Persona represents the beginning of a new era. An era of self exploration based on the work of the philosopher Carl Jung. The Jungian doctorate and professor Murray Stein explains in his book "Map of the soul: persona our many faces" a psychological map in order to know oneself, to know our own soul. When you listen to the first track of the album, "Intro: Persona," when RM (Kim Namjoon) first verse is "Who am I? A question that I've been asking myself for my whole life[12]," it is obvious for Stein that he is talking about "the shadow side that one hides from other.[13]" Now, how is this connected to Twitter? Well, there is a recently converted Jungian analyst ARMY named Laura London who has analyzed lyrics through the Jungian perspective. She also podcasts interviews of different analysts, including Dr. Stein, in order to understand the many layers of Carl Jung's analytical psychology, and how some elements are connected to BTS.

THE BAD

I know it is not all bright and shiny, so as I was getting deeper and deeper into this stan account, I was able to recognize three specific bad habits.

Following the wrong account

It is significant for a new user to know how to identify accounts not to follow. Here are my five cents on this issue.

I understand that you want to gain followers and mutual quickly, but I strongly suggest taking your time to review the profiles of each person that starts following you, or that you want to follow. Here are three steps that will guide you:

1. Read their bio, it will give a gist of their personality or what is the purpose of their account: fan account, fan artists, voting, OT7, solo Stan or OT6, MANAGERARMY, and some fan sites. I will explain each category and their risks later on.
2. Read their older twits. This is a good indication for toxic accounts. Toxic? Well, it means they either tweet or re tweet demeaning messages toward BTS and ARMY. They do it for the clout and gain followers that have the same mindset.
3. Check their following. Verifying who they follow is a great way to know if they are not impersonating and ARMY.

Believe me, there are a lot of accounts that want to start beef with other fandom to redirect our most important asset, our influence to meaningless fights online. If they are not following at least the BTS twitter account, but still twits about them, they are not worth the follow.

Account purposes

Twitter is definitely a huge universe, so I will explain the types of accounts you may encounter and what are the risks to follow them.

- Fan account. Basically, there are different subcategories.

- Soft Stans: they are there for the music, the boys, and any other product BigHit throws quite constantly. For example, photos, collaborations with BT21, Run! BTS, Vlives, Bangtan Bombs, live streaming of concerts, etc.
- Hard Stans: they are there for everything soft stans are plus admiring the boys' physics in a very provocative way. Hey! The boys are gorgeous! Sometimes they tend to over sexualize them, which is reprehensible,

because most of all, they are humans, not pieces of meat. It is okay to thirst, but not to over sexualize.

- Fan artists. There are two types of ARMY that fall in this category. The ones that like to draw are followed by many, because they reimagining BTS in their daily lives, they turn them into fluffy characters, comics, and any other type or form of this art. Then, we have the ones that like to write. First, we need to understand that K-Pop, and actually pop in general, when it comes to boy or girl groups, fans tend to like more some relationships within the members than others. These types of relationships are called, well… ships. It is when the chemistry between two or three is very obvious. These accounts write fictional stories about different ships either in twitter or websites like 'Wattpad' or 'AO3'. The downside for both types of fan artists? Some of them share explicit content, and there are minors using the app without supervision.

- OT7. What is this? A chemical? Not exactly. It means "One True 7." These accounts love the seven members equally. A side note is the definition of a "bias" which is basically your favorite member. You can love all seven and have a bias.

- OT6. Like BTS one said, 7-1=0. BTS are seven, and if you dislike one of them, you are not truly an ARMY. These accounts basically hate one member.
- Solo Stans. They will not phrase is like that per se, but you can tell by scrolling through their timeline that they only tweet or re-tweet ONLY about one member. This category can also blend with the next one or any for that matter.
- MANAGERARMYS. They express their opinions about how, when, with whom BTS should do business. They treat them like if they have no saying in their careers, or even

worse, demand certain things from Big Hit. Let me give you an example: they complain about the amount of official merchandise and collaborations BigHit throw at us every week. It is not like they have a gun in our heads saying we need to buy everything, but they take any excuse to complain about every single detail, as if they can do it better, and have a better insight of the music industry.

- Fan Sites. What they do is follow BTS. There are fan sites that follow BigHit's rules for fans. The most important one is not to approach them off official schedule. On the other hand, there are some fan sites that ignore this golden rule and illegally obtain their plane schedule overseas to take photos inside the airplane, or book a flight in the same plane to be near them. There is a word in Korean for obsessed fans, "sasaeng". They are extremely risky, because Big Hit has a blacklist of Korean and international "fans" to avoid any contact with BTS. So, just imagine how devious they must behave in order to be on that list. They sell photos or other merchandise to maintain that lifestyle. There was one video on YouTube[14] that stated some even prostitute themselves). So yeah, they are a big no-no.

It is not that I forgot to add the famous "trolls" in here, but they are easier to spot. They don't have many followers and the account has been created recently. Those two characteristics are a red flag to stay away from those accounts.

Unpopular Opinion

We all have seen this "unpopular opinion" meme, but there are some profiles that only like to express negative situations, or just reply what others post with the same dark energy. This is a double-edged sword. Let me explain. I am mostly a re-twitter account. I like the way it is, because for me it is hard to convey a message; especially when it comes to clap back.

I know, it seems childish, but there are some instances that a profile needs to be educated. What are some of these topics you need to be aware of? Racism, xenophobia, and misogyny which are the most common ones. I will try to explain how it is intertwined with ARMY twitter on the last chapter.

Here you have two options. If you find a message that covers your thoughts, you can re-tweet. If you haven't seen any, and you would like to add more elements in the conversation, please, speak yourself! Don't be afraid!

Every time you tweet an idea, be ready for any person to misinterpret what you have said. You can either elaborate more by replying, or if you are talking to blatant toxic profile, your time will be wasted, so there will be no need to respond back. There are a lot of people in Twitter that select what they read and understand. That is on them, not you.

If your mutual starts to re-tweet information that is not of your liking, decide where you draw the line. I have un-followed a few of them because they have been re-tweeting information that has been proven wrong or agreeing with any of the three topics that I list earlier.

At the end of the day, do what is best for you. Know your boundaries. Remember that for smaller accounts, it gets trickier, because most want hundreds or thousands of followers. At what cost? This brings me to the final chapter.

THE UGLY

It is of general knowledge that Twitter is a battlefield in disguise. Many people go behind an alias and don't want to deal with the consequences of their written words. It is a shame that it is so hard to make a person accountable for their actions. Having said here is the worst part of twitter.

Mental health

There are some of us that are more prone to take things to heart, and there are others that just like to hurt someone else's feelings. Leah Asmelah from CNN reported through their website on August 2019 that using social media itself it is not harmful, but it takes away time to do productive activities outside and healthy sleeping patterns. These two elements make teenagers, especially girls, to be target of cyber bullying.

In this note, I can say that it is fine to be sad. It is fine to be happy. If you feel depressed all the time by the news or your surroundings, please seek help. Tell one person, a parent, a family member, a friend (on-line or in real life). They will hear you and will try to help you. If any of your options fail, try contact @BTS_AHC, Army Help Center. They will listen to you, and even though they are not Health Care Practitioners, they will try to guide you to seek professional help.

Everyone deserves to be happy, or at least start the path to heal.

Scammers

As there are terrific people on the internet, there are others that scam to whoever allows them to. This is another red flag to avoid, especially on tour season. There are several types of scammers that seek for ARMYs.

- Re-tweets for merchandise. There are some accounts that will offer you BTS merchandise. In exchange, you have to create a tweet saying you need thousands of re-tweets for them to give you an ARMY bomb, an official album, apparel, etc. They ask you for insane number of re-tweets, like five thousand or more. If you are a small account, this is virtually impossible. You will need the help of a big account to do so, and there aren't many big accounts that do this. As a silver lining, there are other types of deals that you can make, for example, with your parents or relatives, and I have seen a lot of people re-tweeting to help an ARMY.

- Scalpers. This is the worst, because they take away possibilities to an actual ARMY to attend a concert. They buy tickets and resell them in a higher amount. I have seen double or triple the original cost. Big Hit has been notified of this several times, but as of today, there have been attempts of lottery tickets in Japan and South Korea. Let's see if they can implement the same measure overseas. My suggestion here is to be patient. A week or two weeks from the concert, there are tickets released because of problems with credit cards, the holders couldn't make it anymore, and you name it. There is a small window of opportunity, but it still there.

- Unauthorized shops. These are normally K-Pop shops online where you can find different unofficial merchandise of different groups. Avoid the ones that have bad reviews or just seem

sketchy. Guard your money, because you may not get it back.
- Fan sites. Yes, here we go again with fan sites. They sell pictures, personalized pins, key chains, etc. Some of them only take advantage of naïve ARMYs and don't send the goodies they promised to. They are called out in the community, but they change the Twitter handle and continue to scam people.

Fan wars

In my opinion, this is inevitable. Let me explain what this is with an example. This past October was

Jimin's birthday, and on that date, a lot of accounts from other fandom started to trend a repugnant hash tag. The beef started early where ARMYs tried to trend positive hash tags, so Jimin couldn't see the other ones. By the way, the fandom trended 20 worldwide hash tags that day at some point. It doesn't matter what the fandom does, there are always some type of backlash. But you know what? Karma is an ARMY. It has been proven time and time again. Fan wars are petty, and we shouldn't give our attention or energy. Besides, with the new Twitter algorithm, every time you interact with an account, it is more likely to appear in other people's timelines and give them more exposure. Do you want to give free PR to this type of people? That's what I thought.

As a special shout out, I would like to include Odie's Twitter handle, @sweetbtstea, in case you want to read special tea.

Misogyny, Xenophobia and Racism

Where do I start? This is a very complex topic, but I will try to make it easier to understand, how can we prevent amplifying the issues, and what can we learn from experiences.

Let's start with misogyny. According to the Merriam-Webster online Dictionary, it is the hatred of women. Kate Manne, in her book "Down Girl: The Logic of Misogyny" explains that misogyny is related to social norms, expectation, and consequences which women have to live under a patriarchal system. Manne is saying that a woman's life should be dictated by a man's perspective. As a clarification, not only men have internalized misogyny, also women. For example, in Latin American countries, it is expected for a young girl to be married between 18 and 25 years old (social norm), to be submissive to her spouse (expectation). Otherwise, a man is entitled to discipline her (consequence) in any way he seems fit, either by inflicting physical or psychological wounds.

On the other hand, xenophobia is *an attitudinal orientation of hostility against non-natives in a given population*, according to UNESCO[15]. The key to understand when to classify comments as xenophobic is the hostile attitude towards either the group itself or the followers. This hostile attitude can come with sarcasm, ill intention comparisons, and ignorance.

Racism has many forms and they happen in any place imaginable. The Australian Human Rights Commission includes in this concept *prejudice, discrimination or hatred directed at someone because of their colour, ethnicity or national original*[16]. It is important to note that it does not only include physical harm, but also micro aggressions like jokes, name-callings, profiling someone according to their last name, and the list goes on.

Having drawn a general perspective to the topics, how it is applied to Twitter then? Well, let me explain with an example that happened on November 2019. The Hollywood Reporter released an interview Seth Abramovitch had with BTS on the second of November[17] where he describes how, in thirteen hours, he learned about the biggest boy band's names and ages, their debut date, their main producer and co-CEO of Big Hit Ent., and a very limited information about K-pop[18]. He also misquotes RM, uses an online translator to understand the name of a typical food, doesn't believe they are genuine by writing "The boys seem to appreciate the gesture, or at least are good at faking it.[19]"

These are only to name a few statements that were not well received, not because the genre or the group is his cup of tea, but the lack of research, racist and xenophobic inference of his writing. A lot of journalists called him out because of lack of professionalism, including Alex Jung, journalist for *Volture* and *New York Magazine*. Seth sent him a direct message to Alex stating that he was always nice to him[20]. The thing is, Alex has never met Seth. This is a prime example of xenophobia, because Seth couldn't even tell two (very different) Asians apart. And the response came a month after the article was released. Jae-Ha Kim, journalist for *Variety, Chicago Tribune, Rolling Stones* among others, describes in a thread[21] the ignorance portrayed by Seth, and making a comparison on how white privilege takes a part of this never-ending story of xenophobia and racism.

It is true that BTS has a majority of female fans, and we are mostly portrayed as raving screaming teenagers. Misogyny comes to play when articles seem to downgrade our passion just because we are mostly females. Let's make a comparison here. Have you seen the lines and camps made before a new iPhone is released? What are the adjectives used? Well, these are people (mostly men) that are eager to grab the newest cellphone. What were the adjectives used when ARMYS camp outside SNL? Crazy teenagers. As we discussed on the first chapter, ARMYs are more than teenagers. We are professionals, students, workers of a wide range of age, skin color, gender identities. There is also an interesting thread about this topic that tackles what I have mentioned before[22]. It mentions other K-pop groups, but the point still stands. Those microaggressions made by, especially, males to downgrade what we enjoy, what we consider music that heals.

CONCLUSION

Twitter can be a scary place for a small account. Apparently, no one is there to help us in case of fan wars or cyber bullying. I am not going to deny that, like in every other community, it has its own bad apples, but they are not the majority. ARMY twitter is a bit different from other communities. We spread love and positivism. We are vocals when we need to, and changes have been made, because we are loud as a collective entity. Just because you are a small account, it doesn't mean, you are less. We are all here because of seven young men that have changed our lives for the better. We have found them, or they have found us. There is saying that BTS come to your life when you needed it the most. I can assure that ARMY will also be by your side in this scary place.

Disclaimer

Efforts have been made to ensure that the information in this book is accurate and complete. However, the author and the publisher do not warrant accuracy of the information, text and graphics contained within the book due to the rapidly changing nature of science, research, known and unknown facts and the Internet. The author and the publisher do not hold any responsibility for errors, omissions or contrary interpretations of the subject matter herein. This book is presented solely for motivational and informational purposes only.

About Authors

Gautam Jain

Gautam Jain lives in San Salvador with his wife and his daughter. He is interested in life and believes in constant self-improvement by being curious towards life, using all means which are in his reach. He loves to travel unexplored places, listening to jazz and sharing his experience to help others.

By being Electronics & Communication Engineer as Academic, worked as Computer Engineer professionally at Infosys where he learned about diversity in life, he understood that life is full of challenges but beautiful. Later completed his Master in Business Administration, he develops interest in various fields. Gautam Jain began his career in Human resources in 2009. In his 10 years of work experience, got chance to work with MNCs and later establish his own consulting business where he provides training completed

HR projects for wide variety of clients. Currently, he is also a University Professor in Central America.

He loves to read other authors about mind, psychology, meditation, entrepreneurship management and communication skills which focus on conscious self-development and take out the best of one's. You can always reach him…. Never hesitate to do so…

Yamilet Rivera

Yamilet Carolina Rivera has an eleven-year experience as an English educator in Central American Universities. She has also been trained as an online tutor by Oxford University. Her educational coordinator position in the GENS foundation from El Salvador has allowed her to create programs for young adults to cultivate job oriented skills. She especially enjoys assisting students to find better ways to approach their personal learning system

References

[1] Caulfield, K. (2019, April 21). BTS Scores Third No. 1 Album on Billboard 200 Chart With 'Map of the Soul: Persona'. Retrieved November 13, 2019, from https://www.billboard.com/articles/columns/chart-beat/8507977/bts-map-of-the-soul-persona-no-1-album-billboard-200-chart.

[2] Chart Data .@BTS_twt's 'Map of the Soul: PERSONA' is now the best-selling album of the year worldwide (via UWC). (2019, April 26). Retrieved from https://twitter.com/chartdata/status/1121573862788030465.

[3] Unterberger, A. (2019, May 2). BTS Show Off Friendship Bracelets, Receive Top Social Artist Trophy On the 2019 BBMAs Red Carpet. Retrieved November 13, 2019, from https://www.billboard.com/articles/news/awards/8509652/bts-2019-bbmas-red-carpet-top-social-artist.

[4] About Us. (n.d.). Retrieved November 13, 2019, from https://www.oneinanarmy.org/about.

[5] Benjamin, J. (2018, November 21). 2016 BTS Song Hits No. 1 on World Digital Song Sales Chart After Fan-Led Initiative. Retrieved November 13, 2019, from https://www.billboard.com/articles/columns/k-town/8486089/bts-2-3-world-digital-song-sales-chart-army-projectbuy23.

[6] iTunes WW#4 - 2!3! (5) https://t.co/FLw8wSCH23. (2018, November 13). Retrieved November 13, 2019, from https://twitter.com/btschartdata/status/1062435352810545158.

[7] #ProjectBuy23 There will be better days!그래도 좋은 날이 앞으로 많기를 pic.twitter.com/SqruPzUQRy. (2018, November 13). Retrieved November 13, 2019, from https://twitter.com/btschartdata/status/1062372592189931521.

[8] Home: BTS Comeback Guidance: A breakdown. (n.d.). Retrieved November 13, 2019, from https://btscomeback.wixsite.com/website.

[9] BTS: Band break UK album and Billboard Hot 100 Chart records - CBBC Newsround. (2019, April 23). Retrieved November 13, 2019, from https://www.bbc.co.uk/newsround/47996730

[10] Kwak, K. (2019, September 5). Big Hit Chief Bang Si-Hyuk, Mastermind Behind BTS, Talks Music, Fans and New Ventures. Retrieved November 13, 2019, from https://variety.com/2019/biz/news/bts-big-hit-bang-si-hyuk-talks-music-fans-1203322755/.

[11] BTS STORYLINE SUMMARY EXPLANATION |

TIMELINE & THEORIES. (2019, April 23). Retrieved November 13, 2019, from https://youtu.be/4Xs-WyzB_zg.

[12] Kim, Namjoon et alli. (2019) "Intro: Persona" (Translator @doolsetbangtan) [Recorded by BTS]. On *Map of the Soul: Persona*. [Digital Audio Copied Recording]. Korea: Big Hit Ent. (2019)

[13] Stein, M., Cruz, L., & Buser, S. (2019). Map of the soul: persona: our many faces. (p-14) Asheville, NC: Chiron Publications.

[14] In Depth Look of Sasaeng Life. (2012, March 23). Retrieved November 13, 2019, from https://www.youtube.com/watch?v=8QEDhh89ZLs.

[15] Xenophobia: United Nations Educational, Scientific and Cultural Organization. (n.d.). Retrieved November 22, 2019, from http://www.unesco.org/new/en/social-and-human-sciences/themes/international-migration/glossary/xenophobia/.

[16] Connie.kwan. (2017, October 4). About racism. Retrieved November 22, 2019, from https://itstopswithme.humanrights.gov.au/about-racism.

[17] Abramovitch, S. (2019, October 2). BTS Is Back: Music's Billion-Dollar Boy Band Takes the Next Step. Retrieved November 22, 2019,

from https://www.hollywoodreporter.com/features/bts-is-back-musics-billion-dollar-boy-band-takes-next-step-1244580.

[18] Idem

[19] Idem

[20] Jung, E. A. (2019, November 2). seth, we have never met pic.twitter.com/QyhDdRa6je. Retrieved November 22, 2019, from https://twitter.com/e_alexjung/status/1190714243030667264.

[21] Jung, E. A. (2019, November 2). seth, we have never met pic.twitter.com/QyhDdRa6je. Retrieved November 22, 2019, from https://twitter.com/e_alexjung/status/1190714243030667264.

[22] Disagree, T. B. (2019, November 14). it's okay to fangirl: how misogyny affects female-centered interests | a thread by @gabyabbyy pic.twitter.com/psLiUkVK2Q. Retrieved November 22, 2019, from https://twitter.com/tybutdisagree/status/1195008051306254337.

www.ingramcontent.com/pod-product-compliance
Lightning Source LLC
Chambersburg PA
CBHW030955240526
45463CB00016B/2676